AF261513

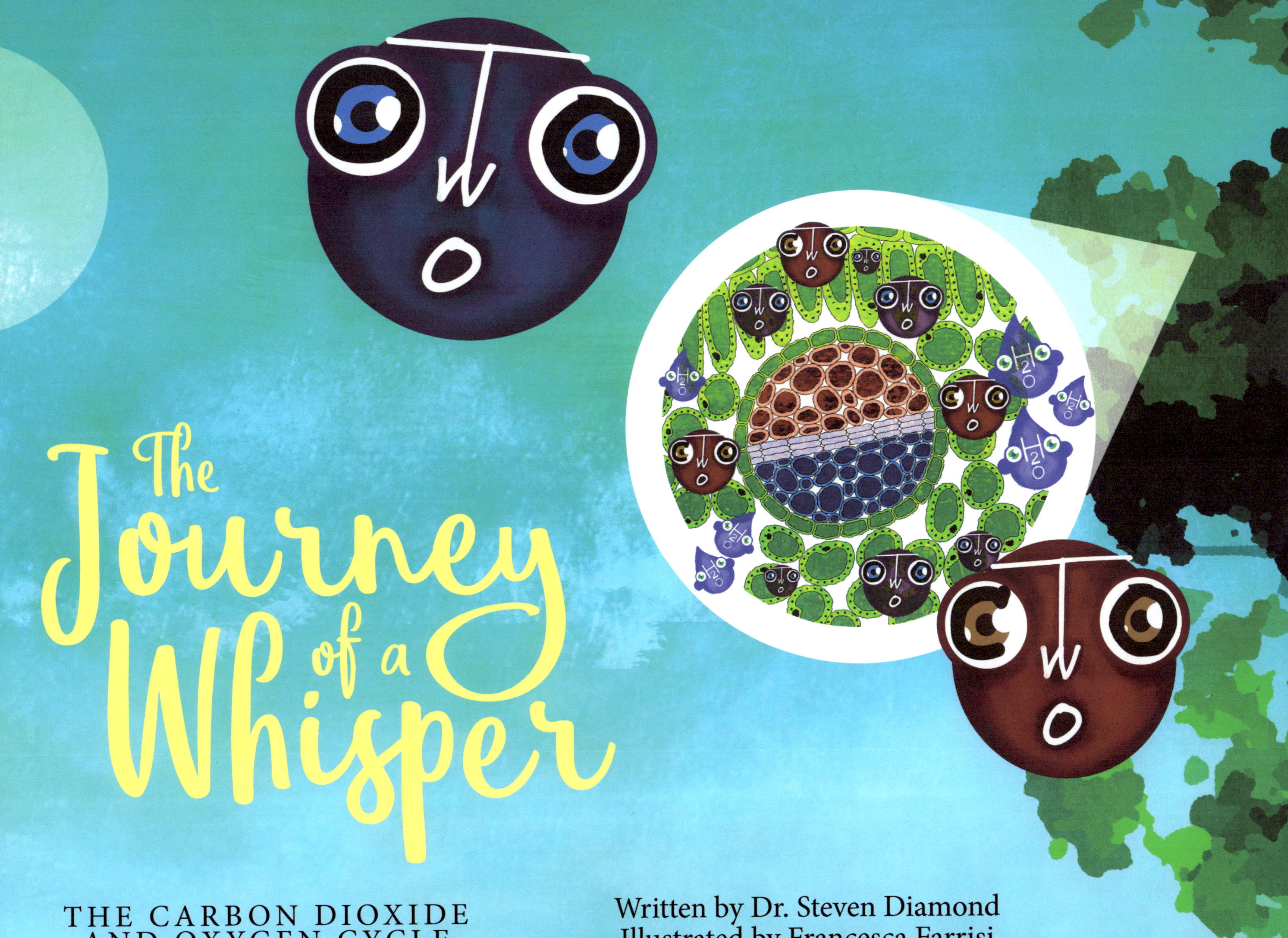

The Journey of a Whisper

THE CARBON DIOXIDE AND OXYGEN CYCLE

Written by Dr. Steven Diamond
Illustrated by Francesca Farrisi

The Journey of a Whisper

THE CARBON DIOXIDE AND OXYGEN CYCLE

WRITTEN BY DR. STEVEN DIAMOND

ILLUSTRATED BY FRANCESCA FARRISI

gatekeeper press

Tampa, Florida

The Journey of a Whisper: The Carbon Dioxide and Oxygen Cycle

Published by Gatekeeper Press
7853 Gunn Hwy., Suite 209
Tampa, FL 33626
www.GatekeeperPress.com

ISBN (hardcover): 9781662943508
ISBN (paperback): 9781662943515
eISBN: 9781662943522

To my wife, Barbara, Children and Grandchildren—
Thanks for your inspiration to write this story.

It's Jose's 6th birthday. His friends celebrate with Jose.

In a whisper, Jose makes a wish, "I wish to go around the world and see all the animals."

During his whisper, Carbon Dioxide (CO2) is exhaled from his mouth.

When he blows out the candles, he
blows the Carbon Dioxide out the door.

CO2 lands on a leaf and through a pore in the leaf it combines with water.

The water (H2O) gets into
the leaf from the ground

when it climbs
up the trunk.

Water and CO2 enter the green chlorophyll in the leaf and make Oxygen (O2) and sugar.

As O2 leaves the leaf into air, sugar is made. Later the sugar will help make maple syrup.

A beautiful red
cardinal breathes
in the O2, flaps his
wings and flies off.

Landing on another tree,
he happily chirps out CO2
from his beak.

Once again oxygen
is made in the leaf.
Jackson, the dog,
breathes it in and
chases a rabbit.

The rabbit leads Jackson to the beach.
Tired, he barks at the rabbit. CO2 comes
from his muzzle.

The green grass uses its chlorophyll to change CO_2 to O_2 for the breaching whale passing by. She sounds and uses the O_2 to cross the Atlantic Ocean.

She surfaces in the Indian Ocean. There along the East Coast of Africa, she blows CO_2 out with her spout of water.

The swaying palm
trees take the
whale's CO2 and
make O2 for a
passing elephant.

After a long trek she raises her trunk and
trumphets her arrival by blowing CO2 out her
trunk. The winds blow her trumpet sound and
CO2 over the snow cap peak of Mt. Kilimanjaro.

Caught in the wind CO2
continues its journey
over Egypt.

And then over India.

...then up, up, up over the Himalayas, the highest mountains in the world. As CO2 travels, it traps Earth's heat.

...and causes snow to melt. The melted snow forms large lakes and rivers.

CO2 continues over the North
Pole enjoying the Northern Lights
caused by Earth's magnetic pole.
It lands in Alaska's tall northern
pine tree forest.

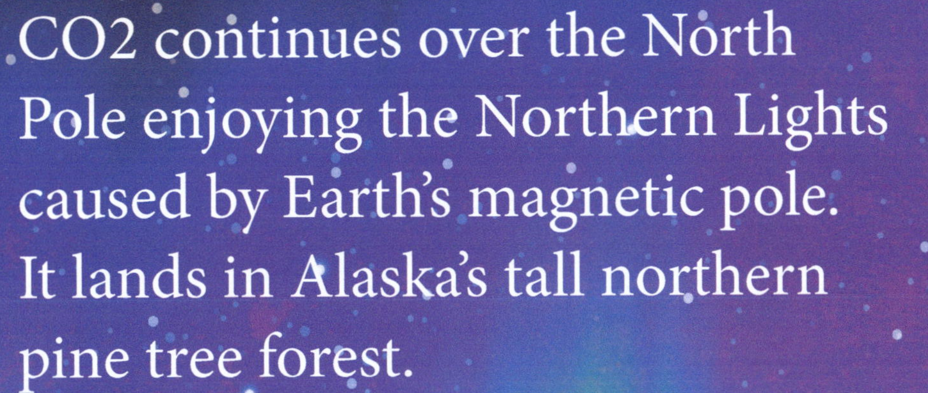

The pine needles make O2 for
the foraging polar bear.

The polar bear bellows a loud hungry growl of CO2 into the air.

The northern winds blow the CO2 south to the great redwood trees of California. The redwoods have made oxygen available for animals for over a 1,000 years.

The O2 is blown eastward.
An eagle breathes it in.

The eagle lets out a high
pitched CO2 squawk.
The CO2 settles on a
tree next to Jose's home.

The O2 is made
once again from the
tree where sugar and
maple syrup was
made last year.

A soft breeze blows O2 into the
house for Jose to take a deep
breath for his 7th birthday.

He inhales O2 and
blows out the candles
once again with CO2.
He makes the same
wish as last year.

Did Jose's wish
come true?

Yes, it did! He visited many strange places and saw animals from all over the world. He did it with the CO2 and O2 cycle!